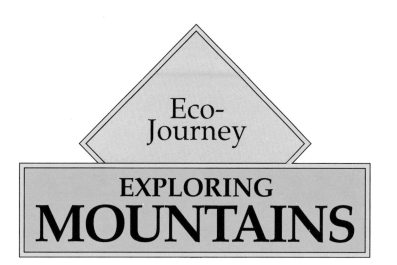

Eco-
Journey

EXPLORING
MOUNTAINS

Eco-Journey

EXPLORING
MOUNTAINS

Barbara J. Behm
Veronica Bonar

Gareth Stevens Publishing
MILWAUKEE

For a free color catalog describing Gareth Stevens' list of high-quality books, call 1-800-341-3569 (USA) or 1-800-461-9120 (Canada).

ISBN 0-8368-1066-X

North American edition first published in 1994 by
Gareth Stevens Publishing
1555 North RiverCenter Drive, Suite 201
Milwaukee, WI 53212, USA

This edition © 1994 by Zoë Books Limited. First produced as *Take a Square of Mountain* © 1992 by Zoë Books Limited, original text © 1992 by Veronica Bonar. Additional end matter © 1994 by Gareth Stevens, Inc. Published in the USA by arrangement with Zoë Books Limited, Winchester, England.

Photographic acknowledgments
The publishers wish to acknowledge, with thanks, the following photographic sources:
t = top *b* = bottom
Cover: Bruce Coleman Ltd.; Title page: Bruce Coleman Ltd.; pp. 6 Robert Harding Picture Library; 7 John Shaw/NHPA; 8 N. A. Callow/NHPA; 9*t*, 9*b*, 10, 11*t*, 11*b*, 12, 13*t*, 13*b*, 14*t*, 14*b*, 15, 16, 17*t*, 17*b*, 18 Bruce Coleman Ltd., 19*t* Laurie Campbell/NHPA; 19*b*, 20, 21*t*, 21*b*, 22, 23*t* Bruce Coleman Ltd.; 23*b* W. S. Paton/NHPA; 24*t*, 24*b*, 25 Bruce Coleman Ltd.; 26 David Woodfall/NHPA; 27*t* John Shaw/NHPA; 27*b* Bruce Coleman Ltd.

Printed in the United States of America

1 2 3 4 5 6 7 8 9 99 98 97 96 95 94

Title page:
A lynx hunts in the winter snow.

CONTENTS

Words that appear in the glossary are printed in **boldface** type the first time they occur in the text.

This is the mountain

A mountain has steep slopes. Its top, or **summit**, is usually very high. Winds blow most of the year on the mountain.

▶ In summer, lower slopes of the mountain are covered by grass and trees. Higher up are cliffs of bare rock.

◀ Cone-bearing trees, such as spruce and fir, grow below the tree line. These trees do not lose their leaves in winter. They can survive the cold and snow.

Trees cannot grow above a level called the **tree line**. It is too cold there for trees.

Early spring

In spring, only the summit of the mountain remains covered in snow. The cliffs below are spotted with plants called **lichens**. In the meadows, beautiful flowers grow.

▶ The lichens on this rock grow well in the pure mountain air. Lichens can survive heat, cold, dampness, and dryness. But they will not grow where the air is polluted.

8

◀ This snow finch fluffs up its feathers to keep warm in the cold wind. The finch's feathers hold a layer of warm air around the bird's body.

▼ The long roots of this purple saxifrage burrow into the cracks in the rocks to find water.

Snow finches spend the winter sheltered below the tree line. In spring, they build nests higher up the mountain, where their young will be safe.

Alpine meadows

Alpine flowers grow quickly in the heat and light of early summer.

▶ In early summer, meadows are carpeted with buttercups, wild strawberries, heather, and more!

◄ An oak eggar moth eats and flies only during the daytime when the sun can warm its body.

▼ A bee drinks the sweet nectar from heather flowers. Bees carry pollen from one flower to the next. After **pollination**, seeds, berries, and other fruits grow on various plants.

The pollen and nectar in the flowers provide food for many bees, butterflies, and other insects. Mountain butterflies often have dark wings. This helps them soak up heat from the sun.

Trees and streams

Cone-bearing trees, or **conifers**, grow just below the tree line. The soil gets deeper and richer lower

▶ An alpine hare changes the color of its fur three times a year. The hare turns white in the autumn, then grayish brown in spring. Shortly afterward, its coat turns brown, making the hare hard to see in the woods.

◀ Salamanders live in cool, moist places, sometimes in water and sometimes on land.

▼ Rocks and boulders beside shady mountain streams are covered by moss and liverwort plants. Liverworts often look like flat, green seaweed growing on the ground.

on the mountains. Broad-leaved trees such as birch and beech grow here. When the snow melts in spring, streams form at the foot of the mountain.

On the mountain

▶ A golden eagle nests on a mountain ledge. It tears off bits of meat it has hunted and gives the food to the eaglets.

▼ Marmots live high in the mountains. They sleep in underground burrows for much of the year to escape the cold.

When warm weather returns, small animals that have spent the winter asleep in underground burrows come alive. Eagles and falcons hunt them from above.

Smaller birds, such as warblers and finches, search for insects or seeds. Caterpillars eat shoots of trees and plants. Each caterpillar will become a **chrysalis** or **cocoon** and then a butterfly or moth.

◄ After wasps mate, the queen finds a sheltered place to sleep for the winter. In spring, she builds a nest by chewing bits of wood to make a paste. With the paste, she builds a structure of papery cells where she will lay her eggs.

Finding a mate

Different types of animals have different ways of attracting a mate. Some dance, and some release an odor.

▶ A female harrier hawk will fly up in the air to the male who drops food he has hunted to her. She catches the prey in mid-air. Sometimes she flies upside down and grabs the food from the claws of the male. The female then takes the food back to the nest to feed the chicks.

◀ A stag's antlers fall off in the spring. New antlers immediately grow to replace them. The new antlers have more branches and points than the previous year's antlers.

Some animals even fight for their mates. Deer mate in autumn. The stag, or male deer, paws the ground and stretches his neck. Then he roars loudly to challenge other stags to fight. The winner mates with the herd's females.

▼ Fern leaves grow from underground stems. At first, the fronds, or leaves, are curled up. As they grow, they uncurl.

Summer insects

During a summer's day, male grasshoppers attract females by making a chirping noise. The noise is made when the males

▶ A bush cricket sips nectar from flowers. Bush crickets are usually more active at night. During the day, they rest on warm stones to absorb heat.

◀ Damselflies live in marshes on the mountain. Their huge eyes allow them to see all around them.

▼ A spider traps an insect in its web. Then the spider injects the prey with a poison, so the victim cannot move.

rub their back legs against their forewings. Tiny insects called mites feed on plants called fungi, other animals, and decaying leaves.

Late summer

Mushrooms and other fungi grow in the damp woods and in meadows in late summer and autumn. In late summer, seeds are

▶ The weasel eats rabbits, hares, lizards, snakes, and birds. It patrols its hunting grounds often to keep other animals out.

◄ Animals do not like to eat the tough stalks and prickly leaves of the carline thistle. In autumn, the flower heads turn silver-white. Thousands of seeds are packed in the flower's center.

▼ Ladybugs cluster together for shelter when the weather gets colder. They return to the same place each winter.

carried away from plants by the wind, by burrs on seeds that hook onto passing animals, and by animals that eat the seeds and later pass them out.

Preparing for winter

During fall, many animals get ready for winter. They fatten themselves in order to survive the cold days when food is scarce.

▶ A female wild boar shows her young how to find food. Boars root, or dig in the ground, for acorns, insects, and larvae.

◀ In autumn, black bears move lower down the mountain, where they can find fruit to eat. The bears' thick long coats protect them from the cold. But in winter, there is not enough food to eat. Then, bears find warm dens under tree roots or in sheltered caves, where they hibernate.

Marmots **hibernate** each winter. When an animal hibernates, its body temperature falls, and its heart rate and breathing slow down. The animal needs very little energy to stay alive.

▼ Stags have fierce battles in the fall. They lock their antlers together and push hard. Eventually, one wins and mates with the females.

Warm and safe winters

Some animals have fur that traps air to help keep them warm. Birds have soft down feathers to keep them warm in winter.

▲ A snowshoe hare changes color from brown to white in the winter. Its feet are broad and covered with hair to keep it from sinking into the deep snow.

▶ In autumn, an ermine's fur turns white, except for the tip of its tail.

24

Some animals molt, or lose, their fur or feathers. Often, the new ones are white, making it hard to see the animal in the snow. This is called **camouflage**.

▲ The ptarmigan molts three times a year, becoming white in winter, brown in spring, and grayish brown in summer.

Winter

▶ Winds blow strongly at the snow-covered summit of the mountain. Plants and animals cannot survive the summit in winter.

In winter, the mountain slopes are covered in snow drifts and ice.

◄ Deer find shelter from the cold winter winds below the tree line. They scrape through the snow to find moss and lichen to eat.

▼ Wet snow has fallen on the branches of these conifers. The pointed shape of the trees will allow the snow to slide off the branches.

Animals live below the tree line in winter to keep out of the cold wind. Mice, voles, and shrews live in tunnels beneath the snow, away from the cold wind. The air underneath the icy top layer of snow is warmer, moist, and calm.

More Books to Read

First Look at Mountains. Susan Baker (Gareth Stevens)

Ming Lo Moves the Mountain. Arnold Lobel (Scholastic)

Mountain Homes. Althea (Cambridge University Press)

Mountains and Volcanoes. Eileen Curran (Troll)

The Time a Cloud Came into the Cabin. Jacquelyn Smyers (Very Idea)

A Walk Up a Mountain. Caroline Arnold (Silver Press)

Videotapes

Call or visit your local library to see if these videotapes are available for your viewing.

Mountain Regions, about Yellowstone National Park. (Encyclopedia Britannica Educational Corporation)

Mountains and Mountain Building. A Natural Phenomenon Series (Journal Films)

Places to Write

For information regarding nature centers in your area, contact:
National Audubon Society
700 Broadway
New York, NY 10003

For more information regarding mountains and wildlife, contact:

U.S. Department of
 Agriculture
Forest Service/Public
 Affairs Office
Publications
Auditors Building 2 Central
201 Fourteenth Street, S.W.
Washington D.C. 20250

Internal Ministry of the
 Environment
Public Information Center
First Floor
135 St. Clair Avenue, West
Toronto, Ontario M4V 1P5

Interesting Facts

1. Water drains away quickly on steep mountain slopes, so many mountain plants have spongy leaves to hold water.

2. Fine hairs on mountain plant stems and leaves stop the wind from drying up any moisture.

3. Pollen and nectar in flowers provide food for bees, butterflies, and many other insects.

4. Mountain insects, such as butterflies and bees, stay as close to plants as possible, so that the wind will not blow them away.

5. Leafcutter bees cut off bits of leaves with their jaws to make their nest.

6. The dark color of alpine salamanders helps the animals absorb, or take in, heat from the sun to warm their bodies.

7. In summer, the marmot digs a maze of tunnels in which to run and hide if another animal is hunting it.

8. Male black grouse dance at dawn to attract females. The males gurgle, spread out their tails, droop their wings, and jump in the air to attract a mate.

9. Some mountains are formed when volcanoes spew up lava and rock that hardens as it cools.

10. Mountain sickness gives people shortness of breath and sometimes nausea, headaches, and even nosebleeds. This is because of the lack of oxygen at high altitudes.

Glossary

alpine: having to do with high mountains.

camouflage: the coloring or shape of an animal that makes it difficult to see against its background.

chrysalis: a resting stage in the life of a butterfly, between the caterpillar and the adult insect.

cocoon: a resting stage in the life of a moth, between the caterpillar and the adult insect.

conifers: trees that bear seeds in the form of cones.

hibernate: to spend winter in a state of rest.

lichens: two plants that live together, a fungus and an alga, and help each other survive.

pollination: the carrying of pollen from a male to a female flower.

summit: the top of a mountain.

tree line: the level on a mountain above which trees do not grow.

Index